Jolee was going camping over the July 4th weekend. She had never been camping before.

"I know I am going to love it!" she thought.

Mom, Dad, Jolee, and Josh were off! Jolee checked her knapsack. Did she omit anything? She had put in a flashlight, notebook, and pen. At night, she would write about her day of camping.

"Are you ready to rough it, kids?" asked Dad. "We won't be staying in a hotel!"

"Dad likes to joke," said Mom. "The campgrounds are really not that rough."

Soon they saw the sign for the campgrounds. "We are here!" shouted Jolee and Josh.

Jolee and Josh helped put up
the tent.

"You kids are tough!" said Dad.
"I should have known you would be!"

Before lunch, they went for a swim. Later on, they took a short hike. Then they went back to their campsite.

At bedtime Mom, Dad, and Josh
went to sleep. But Jolee was still
awake. Now she wanted to write about
the events that day.

"These hot dogs are superb!" said
Mom. "Dad cooked enough for an
army!"